Animal Marvels

Finding Food

Please visit our website at: **www.garethstevens.com**
**For a free color catalog describing Gareth Stevens'
list of high-quality books and multimedia programs,
call 1-800-542-2595 (USA) or 1-800-461-9120 (Canada).
Gareth Stevens Publishing's Fax: (414) 332-3567.**

Library of Congress Cataloging-in-Publication Data available upon request from publisher. Fax: (414) 336-0157 for the attention of the Publishing Records Department.

ISBN 0-8368-2815-1

This North American edition first published in 2001 by
Gareth Stevens Publishing
330 West Olive Street, Suite 100
MIlwaukee, WI 53212 USA

© QA International, 2000

Created and produced as *So Many Ways to Eat* by

QA INTERNATIONAL
329 rue de la Commune Ouest, 3ᵉ étage
Montréal, Québec
Canada H2Y 2E1
Tel.: (514) 499-3000 Fax: (514) 499-3010
www.qa-international.com

Printed in Canada

1 2 3 4 5 6 7 8 9 05 04 03 02 01

Gareth Stevens Publishing
A WORLD ALMANAC EDUCATION GROUP COMPANY

Food for life

Finding enough to eat is a matter of life and death. Many animals spend almost every minute of their time just finding food. Animals need food to grow, to move around, or just to stay alive. Mother Nature's menu includes many foods: grasses, fruit, seeds, roots, nectar, and meat. Animals sometimes have to work very hard and use special methods to get their fill.

Help in digestion

Some plants are hard to digest. Some animals, such as cows, deer, and sheep, have solved this problem. First, they chew and swallow their food. It is then partly digested in the stomach. Then the food is brought back up to the mouth. It is chewed and swallowed again. The plants are digested while they are inside different parts of the stomach.

2

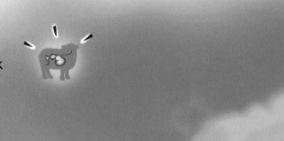

sheep

A powerful meat eater

The tiger is the biggest of the big cats. It is very still as it waits for its next meal. When its victim comes close enough, the tiger charges. It sinks its teeth into the neck of the animal. Death is very quick. The tiger roars! Then it carries its meal to a place where it can eat in peace. Tigers are fierce animals, but they miss their prey 95 percent of the time!

Siberian tiger

The ocean strainer

What animal can eat nearly 9 tons (8,000 kg) of food per day? The giant blue whale can! This whale swallows huge amounts of seawater. The blue whale's mouth has hundreds of tiny plates known as baleen. The baleen strains the water for millions of tiny crustaceans called krill. The krill is this baleen whale's main food supply.

blue whale

3

Flower feeders

Many insects, bats, small rodents, and birds drink nectar. The ruby-throated hummingbird is only 3.5 inches (9 centimeters) long. It hovers in front of a flower and sticks its beak into the flower. The hummingbird unfurls its long tongue, which acts like a straw so the bird can suck the nectar from the flower. This hungry bird feeds 15 times an hour!

ruby-throated hummingbird

Are you curious?

When sheep in a flock sense danger, they huddle together. The sheep with the most experience know enough to get out of harm's way. The rest follow. This is probably why we talk about people who "follow one another around like sheep."

Predators

Predators run, swim, leap, and dive. They set records that the greatest Olympic champions cannot break! They hunt alone or in packs. They set traps and use tools — and they use special weapons. Predators are smart and have supersharp senses. Most can find their prey easily. Foxes, for example, use their sensitive hearing. Dogs use their keen sense of smell. Many birds, such as the eagle, use their amazing eyesight.

A deadly hug

The largest African snake is the rock python. This huge snake can grow to 26 feet (8 meters) long. It shoots up like a spring and tightens itself like a vise around its victim until the victim is suffocated. Then the rock python loosens its grip and starts eating its catch. It can eat an antelope in no time at all!

4

African rock python

A champion dive bomber

The peregrine falcon is about the size of a crow. When it spots its prey from high in the sky, it dives headfirst. It can reach speeds of nearly 200 miles (320 km) per hour! The victim dies from the force of the blow from the talons (claws). The peregrine falcon can kill prey as large as geese, herons, or cranes!

peregrine falcon

cheetah

Speed demon

The cheetah watches its prey from afar. Then it sneaks a little closer. Suddenly, it heads straight for its target. It throws the animal off balance with one stroke of its paw. Then it grabs its prey by the throat and strangles it. The cheetah is a champion runner. It can reach speeds of 70 miles (112 km) per hour! It eats only about 7 pounds (3 kg) of meat per day.

5

A dreaded enemy

The great white shark is a frightening animal. It can grow up to 30 feet (9 m) long. Its jaws are full of teeth that are sharp as knives. This shark attacks its victims from behind and quickly kills its prey. The great white shark can snatch 20 pounds (9 kg) of flesh in a single bite!

great white shark

Are you curious?

Boas and pythons are constrictors. These snakes kill their prey by hugging them tightly. Some pythons grow to be over 30 feet (9 m) long. Their bodies can be more than 15 inches (nearly 40 cm) around. They can weigh over 220 pounds (100 kg)!

Trap setters

Trap setters are tricky hunters. Some rely on camouflage and take on the form or colors of the places they live. Others set clever traps. These animals often get what they want with very little effort, and they succeed in catching prey almost all the time. Everywhere in nature we can see examples of camouflage and other animal traps.

A "praying" insect

The mantid is sometimes called a praying mantis. Mantids are masters of camouflage. They look like the plant on which they live. Some species look like orchids. Others have limbs that look like petals. This beautiful insect unfolds its pincer-shaped front limbs and sinks its sharp spines into the body of its victim.

flower mantid

6

Are you curious?

The front legs of the praying mantis are often joined together as if in prayer. But the mantid is not praying! It is one of nature's most murderous insects. Certain tropical mantids can even capture birds and lizards!

A cowboy spider

The bolas spider spins a thread with a sticky ball on the end. The ball smells like a female moth and attracts male moths looking for a mate. When a moth comes close, the spider twirls the thread like a cowboy twirling a lasso! The spider then captures the poor moth and makes a meal of it.

bolas spider

European anglerfish

A fishing fish

The anglerfish carries a deadly trap on top of its body. It has three spines on its back. One of them is right next to the mouth. This spine is topped by a small piece of flesh that is like a makeshift fishing rod! When the angler waves its rod, its prey mistakes it for a plump little worm. That is when the huge mouth opens up and sucks in the prey!

7

A lion among ants

The ant lion larva uses its tail like a shovel to build a pit in dry, sandy soil. When the pit is finished, the larva buries itself in the bottom. When ants and other walking insects fall into the pit, the ant lion kills them with its long, sharp jaws.

ant lion larva

Hunting parties

What do sharks, wolves, and hyenas have in common? They all hunt in groups. These predators work with other members of their species, especially if their prey is bigger and faster than they are. The hunters spread out and surround their victim so it can't escape. A single lion can kill a gazelle only 15 percent of the time. A group of two or three slays its victim 32 percent of the time! The size of the hunting party depends on the prey. Two or three hyenas can take down a gazelle. It takes about a dozen to kill a zebra!

Fishing buddies

White pelicans have a clever way of hunting. First they form a semicircle around a school of fish. Then the pelicans strike the water with their wings and put their beaks in the water. They force their prey into the center of the group. The white pelican takes in more than 2 pounds (1 kg) of food each day.

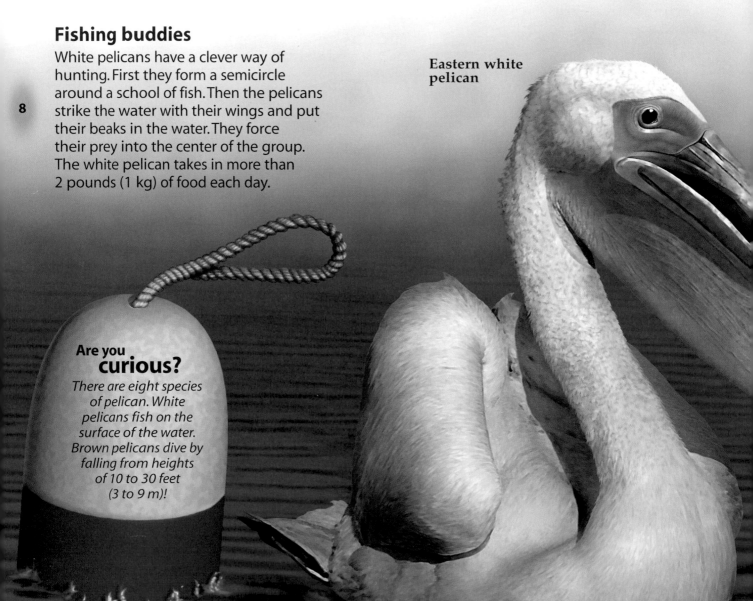

Eastern white pelican

Are you curious?
There are eight species of pelican. White pelicans fish on the surface of the water. Brown pelicans dive by falling from heights of 10 to 30 feet (3 to 9 m)!

Beware of dog!

African hunting dogs are the best hunters in Africa. They are lightning fast and have great endurance. These dogs have mastered the art of hunting in a pack. In fact, their success rate as hunters is 80 to 90 percent. Most of their attacks take just a couple of minutes! These hunting dogs kill gazelles, gnus, and zebras.

African hunting dog

legionary ant

Armies of ants

Millions of hungry legionary, or army, ants march together through tropical forests. Their columns can be up to 8 inches (20 cm) wide and up to 1,000 yards (900 m) long. They eat every living thing in their path. They are even said to have devoured a chained leopard in just a few hours!

Teamwork

Killer whales live and hunt in large groups. To capture schools of salmon, they form a U-shaped hunting party and force the fish to head for shore. Once encircled, the salmon are eaten one by one. These whales can grow to be 33 feet (10 m) long. They can swim as fast as 35 miles (56 km) per hour.

killer whale

Helping each other

Many parasitic animals feed off others in a way that harms the host. However, many animals live together peacefully and exchange favors. To maintain life without much effort, some animals help others. In exchange, they get to feed on the food their companion provides. These relationships are called "symbiotic." They are sometimes so helpful that the animals cannot live without each other.

Cleaning done cheap!

Some small fish set up "cleaning stations" in coral reefs. These fish, called cleaner wrasse, clean out the gills, mouths, and fins of large fish. As they clean, they feast on the parasites that clutter these body parts. Wrasse stations are so busy that several fish may be lined up waiting their turn. The busy cleaners can serve more than 300 customers in six hours!

leopard moray eel

Pacific cleaner wrasse

A walking food supply

The oxpecker can always find something to eat! It snaps its beak like scissors to remove the ticks and parasites that attach themselves to the coats of large African mammals. The oxpecker spends most of its time clinging firmly to the fur on the back, legs, and even the belly of buffaloes, zebras, giraffes, and antelopes.

yellow-billed oxpecker

Egyptian plover

A tooth-picking bird

With its mouth wide open, the dreaded Nile crocodile waits for its teeth to be picked. A small, brave bird such as the Egyptian plover slips between the crocodile's sharp teeth. It feasts on the pieces of food that are stuck there. At the first sign of danger, the bird makes piercing cries to let the crocodile know it's time to slide back under the water!

sponge crab

Protective headgear

Many crabs use camouflage to change their looks. The sponge crab attaches a piece of sponge to its shell. As the sponge grows, it covers the shell. The sponge hides the crab and also tastes bad. Predators won't eat it. In exchange for helping the crab, the sponge is always moved to new feeding grounds.

Are you curious?

Some small fish are the same color as the cleaner wrasse. They also mimic the behavior of that fish. Once they get close to a fooled customer, they start tearing off pieces of its fins and skin!

The right tools

Animals need sharp senses and clever ways of hunting to find food. They also need tools to help get the food into their mouths and swallow it. Insectivores (insect eaters) have tiny, pointed teeth. Herbivores (plant eaters) have large, flat molars. Carnivores (meat eaters) have fangs as sharp as swords. Termite eaters have long, sticky tongues. Whether it eats plants, meat, or insects, each animal has the perfect tools.

An exploring finger

As it travels through the forests of Madagascar, the aye-aye listens for sounds of life under the bark of the trees. Its powerful senses can detect larvae hidden inside the trees. It then puts its long, thin third finger to work. It explores the cracks in the bark, captures the larvae, and puts them in its mouth.

aye-aye

A meat-eating shellfish

This pretty murex is a carnivore. It is a shellfish that feeds on spineless animals of tropical seas. It secretes an acid that makes a hole in the shell of a fellow mollusk. Then it sticks a tube into the hole. Tiny teeth inside the tube quickly devour the victim.

dye murex

A sticky trap

The toad uses its tongue to catch its prey. Its tongue is coated with a sticky mucus. When an insect comes close, the toad shoots out its tongue, catches the insect, and swallows it in less than a second! To help the toad swallow, muscles force its eyeballs into its mouth. The eyeballs are used to push the food down its throat.

common toad

13

A helpful tool

No shellfish is safe from the oyster catcher! It has a long, bright-orange beak that can open cockles and mussels. It pushes the sharp tip of its beak into the muscles that keep the shell closed. Then it twists its beak to spread the two sides of the shell apart. The oyster catcher's meal waits inside!

oyster catcher

Are you curious?
The aye-aye is the last surviving member of a very old branch of the primate family. This animal is now very rare on the island of Madagascar. Its natural habitat has gradually disappeared because human beings have destroyed the forests.

Special tricks

All prey cannot be swallowed easily. Some animals protect their tender bodies with hard shells. Many hide in trees. Others rely on speed to escape their enemies. But predators do not give up either. Many of them have developed ways to outsmart even the trickiest prey!

A garden dinner table

Broken snail shells are all that's left of a song thrush's meal. The thrush takes a snail in its beak and throws it several times against a stone or a rock. This breaks the shell and loosens the muscle that keeps the animal in its home. Then it's time for a delicious meal!

song thrush

The egg thrower

Each species of mongoose has its own way of cracking a tough egg. The zebra mongoose crouches with the egg in its front paws and throws it back between its hind legs. The egg smashes against a rock. This trick also comes in handy when the mongoose wants to eat a tasty snail, a crab, or a giant pill bug.

zebra mongoose

leopard seal

The sneaky seal

The Adélie penguin is a fast swimmer, but it cannot escape the leopard seal. This large seal hides under a block of ice and waits until a penguin dives into the water. Then it snaps up the penguin. A single leopard seal can eat as many as 36 of these pengins in one day! Leopard seals also eat crabeater seals, seabirds, and fish.

crow

15

Seaside meals

The Northwestern crow eats snails, mussels, and other seaside mollusks. As it patrols the northern Pacific coastline, the crow picks up mollusks. It drops them on the rocky beaches from a height of about 16 feet (5 m). The shells break open to expose the tender bodies inside, which the crow immediately gobbles up.

Are you curious?

The song thrush is famous for its beautiful singing. It is also known for its talent as an impersonator. It imitates the songs of many other bird species.

Tool users

So you think spoons and knives were the first utensils ever invented? Wrong! Since long before human beings appeared on Earth, animals have used tools for flushing out food. They use twigs, pieces of wood, stones and rocks, and even the ground itself. Some birds even stab their victims on twigs or cactus spines!

A delicious treat

The chimpanzee has mastered many tools. One of its favorites is the stripped tree branch it uses to invade termite and ant nests. As soon as the stick enters the nest, thousands of termites attack it. When the chimp pulls out the stick, it is covered with insects, which the chimp then eats!

common chimpanzee

A stone tool

Sea otters of the Commander Islands feed on sea urchins, mussels, and crabs. The otters bring up a flat stone from the seabed to the surface. They float on their backs with the stone on their abdomens. Then they grab their prey and strike it against the stone until its hard shell breaks apart.

sea otter

woodpecker finch

Larvae on a stick

The woodpecker finch uses a strong, thin twig or cactus needle to capture its food. Holding this tool in its beak, the finch pokes into the holes and cracks in tree bark. It is looking for the insect larvae that live there. When it finds them, it either skewers them or pushes them toward an opening in the bark.

Egyptian vulture

A runny feast

The Egyptian vulture loves hard-shelled oyster eggs. When it spots a nest of them, the greedy bird picks up a large pebble in its beak. It drops the pebble on the eggs several times until the shells break. The vulture likes the eggs so much that it will run up to 3 miles (5 km) to find the right stone.

Are you curious?

Chimpanzees survive mainly on fruit and insects. From time to time, however, they also kill animals and eat meat. The animals they feed on include baboons, pigs, and antelopes.

Powerful weapons

Many animals are equipped with weapons that can shake, paralyze, stun, and kill. The ones that have these weapons can usually capture their prey easily. Many animals are equipped with venom, an excellent weapon. Bees, wasps, and certain snakes all have poisons that can kill in just a few seconds. Other animals, such as electric eels, can deliver stunning shocks.

Dangerous droplets

The archer fish is a hunter with a weapon that is deadly and accurate. The fish holds its body in a vertical position just below the surface of the water and shoots a stream of water droplets. The stream hits the small insects resting on vegetation above the water! The victims fall into the water and are immediately eaten.

archer fish

A deadly poison

Vipers have two hollow fangs that carry a poison produced in special glands. When it opens its mouth to gobble up its victim, the viper swings the fangs forward. It sinks the fangs into the body of its prey and shoots its dreadful poison into it. Then it swallows the animal whole.

aspic viper

A deadly trap

Sea anemones are colorful creatures that look like flowers. They attack shrimp and small fish that brush against their tentacles. The tentacles contain cells with poison threads that pierce the skin of the victim and paralyze it. They then bring the animal toward the mouth of the anemone.

pink-tipped anemone

A shocking catfish

The skin of the electric catfish contains an organ that can send out sharp shocks. This African hunter stuns fish and invertebrates by bombarding them with a series of electrical impulses. These strong charges can deliver up to 350 volts of electricity!

electric catfish

Are you curious?

A groove runs across the palate of the archer fish. When the fish presses its tongue against this groove, a narrow canal is formed. The fish then creates pressure that shoots a strong stream of water droplets from its mouth.

Planning ahead

To survive winter food shortages or prepare for hard times, some animals store up food. Their pantries can be simple holes, special burrows, or rooms attached to their main dwelling. Many animals store fresh or leftover food. Wolves and coyotes bury their leftovers underground. Big cats hide theirs under twigs, branches, and leaves.

A pantry in the sky

The panther's open-air pantry is safe from prowlers because it is high off the ground in the fork of a tree. The big cat carries its catch between its powerful jaws as high as 30 feet (9 m)! This task takes so much energy that the panther may have to wait up to half an hour before it can summon up the strength to eat its prey.

panther

Ants that eat their fill

Honey-pot ants cling to the ceiling of their nest and stay perfectly still. During the mild season, worker ants cram these insects full of sap. Their stomachs become as big as green peas. When food is scarce, the other ants in the colony visit these hanging "honey pots." The honey-pot ants release their supply of sweet liquid to feed the other members of the colony.

honey-pot ant

great grey shrike

A thorny death

The great grey shrike is just a bit larger than a sparrow. This tiny torturer sends its prey to a very painful death. It hunts insects, spiders, small rodents, reptiles, and amphibians. It impales them on the needles of thorny bushes. With a well-stocked pantry, the shrike can survive food shortages. Using its hooked beak, it can carve up its food with ease.

Woodpecker pantries

The trees, posts, fences, and wooden walls of the southwestern United States are riddled with little holes. The acorn woodpecker drills these holes to serve as food lockers. It uses its strong beak to force acorns into the holes for a winter food supply.

acorn woodpecker

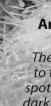

Are you curious?
The black panther belongs to the same species as the spotted panther. Its beautiful dark coat is different from the spotted coat. The animals themselves are no more different than people with dark skin or light skin.

Picky eaters

Most animals prefer certain foods, but a few are really choosy about what they eat. Some animals pass up most of the food they find. These animals inspect every morsel that will end up in their stomachs. Because they are such picky eaters, they lead risky lives. If their favorite food is not available, they may not survive.

A delicate stomach

The daily menu of the koala consists of 1 pound (0.5 kg) of eucalyptus leaves. Using its sense of smell to figure out which leaves it can eat, the koala chooses its food with great care. It never eats the young shoots, which contain a deadly poison.

koala

green discus

Food that sticks to the skin

Green discus parents make their own baby food — a mucus secreted by their skin. A few days after their birth, the baby fish attach themselves to the skin of one of their parents and feed on this mucus. A few weeks later, the young fish can start eating prey they hunt for themselves.

Blood on the menu

Vampire bats have nothing in common with the bloodthirsty monsters that share their name – except their taste in food. They rest in the dark during the day. At night, they come out to drink the blood of sleeping animals. These bats use their teeth to pierce the skin of their victims and drink the warm blood that flows from the wound.

common vampire bat

giant anteater

An eater of ants

The anteater feeds on very tiny prey. The tongue of a large anteater is 24 inches (61 cm) long. The anteater inserts its tongue into an opening in an ant or termite nest. The tongue is coated with sticky saliva. It darts in and out to catch the little insects. The anteater consumes at least 30,000 ants per day!

Are you
curious?

At birth, a koala measures less than 1 inch (2.5 cm). Like all marsupials, after its birth it completes its development in a pouch on its mother's belly. It remains there for five to seven months.

Garbage collectors

Scavengers search for meals left by others. They don't risk their lives or use much energy. They eat the ready-made meals that are handy. The food is often of poor quality, but scavengers don't have to catch their own food. They don't risk the injury or death that the hunters risk. These garbage collectors serve a very useful purpose. They clean up the planet's environment.

turkey vulture

A bird that preys on the dead

Using its hooked beak to tear off strips of flesh, the turkey vulture feeds on dead animals. This bird has excellent eyesight. It can spot carrion (dead and rotting flesh) from a great distance. The vulture can live on such bad food because juices in its stomach help make the food safe.

24

Grave-digging insects

Sexton beetles work in pairs. Together, these very strong insects prepare their food supply. They slide beneath the dead body of a small mammal or a bird and dig a pit until the animal is buried. They remove the feathers or hair and form the flesh into a ball that will be used to feed their larvae.

sexton beetle

common whelk

Deep-sea scavenger

The whelk eats live crabs, worms, and mollusks. It also eats the flesh of dead animals – as long as it's fresh! It uses its long, pinkish tongue to scoop up the bodies of dead crustaceans and mollusks that litter the ocean floor. This scavenger therefore helps rid the oceans of debris that could cause pollution.

25

Top-notch scavengers

The brown hyena roams at night in search of food. It eats insects, small mammals, eggs, and fruit. It also eats the remains of dead animals left behind by others. The hyena will eat anything! When it finishes its meal, whatever it cannot digest comes back out of its mouth in the form of pellets.

brown hyena

Are you curious?

Because of their feeding habits, vultures' heads are completely bald. This allows the vulture to put its head inside the bodies of dead animals without soiling its feathers!

A map of where they live

More fun facts

ANIMAL DIETS		
Animal	**Diet**	**Examples**
Carnivore	meat	boas, pythons, cobras, owls, dogs, cougars, ermines, lions, wolves, martens, polar bears, eagles
Insectivore	insects	spiders, chameleons, nightjars, swallows, shrews, moles, armadillos, ladybirds, anteaters, dragonflies, praying mantids
Herbivore	grass and other plants	cows, bison, antelopes, elephants, sea urchins, caterpillars, tadpoles, rabbits, boars, gazelles, giraffes, manatees
Frugivore	fruit	monkeys, robins, dormice, fruit bats
Granivore	seeds and grain	pigeons, hens, parrots, chaffinches
Piscivore	fish	ospreys, gannets, herons, penguins, kingfishers, auks, Mexican bulldog bat, land otters
Nectarivore	nectar	honey mouse, hummingbirds, bees, butterflies, mongoose lemurs
Omnivore	a large variety of food (fruit, seeds, grain, insects, meat, honey, eggs, fish)	ants, rats, yellow ground squirrels, chimpanzees, raccoons, red foxes, black bears
Detritivore	dead animals and plants	cave crabs, dung beetles, wood lice (crustaceans), blue-bottle and green-bottle flies, carps, crows, seagulls, vultures

WHO AM I?

1. This primate can eat up to 50 bananas one after another.
2. This caterpillar with a huge appetite eats 86,000 times its own weight in leaves during the two days after it emerges from its egg.
3. This huge reptile not only swallows its prey whole, it also eats stones to help it crush and digest food in its stomach.
4. This lover of ants and termites eats so noisily that it can be heard from 219 yards (200 m) away.
5. This animal digests its meal outside its body in a stomach that comes out of its mouth.
6. These little insects form and bury balls of manure on which they lay their eggs and feed their young.
7. In schools of several hundred, these 12-inch- (30-cm-) long fish can devour enormous prey with lightning speed.
8. This insect weighs only 0.12 ounces (3.5 g) and consumes the equivalent of its own weight in food every day. Sometimes forming swarms of several billion individuals, it does a great deal of damage to crops.
9. Although this little animal weighs only between 1.4 and 2.1 ounces (40 and 60 g), it still manages to consume 6.3 ounces (180 g) of food per day.
10. This insect "tastes" food using chemical receptors in its feet that determine how much salt, sugar, and water it contains.

*Answers at the end of the glossary (p. 31).

WHAT'S ON THE MENU?

Animals	What they eat
Aphid	Plant sap
Barracuda	Fish, young seabirds
Beaver	Bark
Common squirrel	Wild fruit, hazelnuts, acorns, chestnuts
Cottontail rabbits	Grasses, grain, vegetables, bulbs, twigs, bark
Crowned eagle	Antelopes, monkeys, hyraxes, hornbills
Domestic hen	Seeds
Dragonfly	Mosquitoes, flies, bees, small butterflies
Dung-beetle	Excrement
Earthworm	Humus (decomposing organic matter)
European goldfinch	Seeds, insects
Giant panda	Bamboo stems
Giraffe	Acacia leaves, buds, fruit
Grey flycatcher	Butterflies, bees, beetles, dragonflies
Gull herring	Fish, eggs, chicks, small mammals, refuse
Hippopotamus	Plants
Honey mouse	Nectar, pollen, sometimes insects
Honeybee	Nectar, pollen
Ladybird	Aphids
Lion	Gnus, zebras, gazelles, buffalo, warthogs, hartebeests
Marine iguana	Seaweed
Moray eel	Fish, large crustaceans, octopuses
Mulberry-silkworm caterpillar	Mulberry leaves
Naked mole rat	Tubers, roots
Nautilus	Small crustaceans, hermit crabs, small crabs, spiny lobsters, fish
Parrot	Nuts
Piranha	Fish
Polar bear	Seals, reindeer, fish, rodents
Prezwalski's horse	Grasses
Puma	Mice, rabbits, hares, squirrels, deer, wolves, small mammals, large mammals
Robber crab	Coconuts
Rock dove	Seeds, grasses, snails, mollusks
Snail	Leaves, stems
Spanish lynx	Rabbits, otters, rodents, young boars, birds
Sperm whale	Octopuses, squid, fish, sharks, rays
Swordfish	Mackerel, herring, squid, young tuna and flying fish
Termites	Wood
Weasel	Insects, mollusks, lizards, birds, small rodents
White stork	Insects, small rodents, amphibians, reptiles, earthworms
White-butterfly caterpillar	Cabbage leaves
Wolverine	Reindeer, rodents, hares, wasps

For your information

domestic sheep

size	up to 3.5 feet (1 m)
weight	100 to 350 pounds (45 to 159 kg)
distribution	raised in many countries
habitat	pasture
diet	herbaceous plants
reproduction	1 lamb; 144- to 153-day gestation period
predators	stray dogs, wolves

class | Mammals
order | Artiodactyla
family | Bovidae

African rock python

size	10 to 22 feet (3 to 7 m) long
weight	33 to 66 pounds (15 to 30 kg)
distribution	sub-Saharan Africa
habitat	woodlands, savannahs, scrub
diet	rodents, birds, gazelles, impalas
reproduction	between 50 and 100 eggs; 2- to 3-month incubation period
predators	crocodiles, hyenas, tigers
life span	20 years in captivity

class | Reptiles
order | Sqaumata
family | Boidae

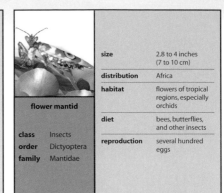

flower mantid

size	2.8 to 4 inches (7 to 10 cm)
distribution	Africa
habitat	flowers of tropical regions, especially orchids
diet	bees, butterflies, and other insects
reproduction	several hundred eggs

class | Insects
order | Dictyoptera
family | Mantidae

---

eastern white pelican

size	4.6 to 6 feet (1.4 to 1.8 m) wingspan: 9 to 12 feet (2.7 to 3.6 m)
weight	22 to 24 pounds (10 to 11 kg)
distribution	Europe, Near East, Asia, India
habitat	lakes, lagoons
diet	fish
reproduction	2 eggs; 30- to 35-day incubation period
predators	sea lions, sharks

class | Birds
order | Pelicani-formes
family | Pelicanidae

Pacific cleaner wrasse

size	about 4 inches (10 cm) long
distribution	tropical oceans and seas
habitat	lagoons, coral-rich areas
diet	external, oral, and gill parasites of fish

class | Fish
order | Perciformes
family | Labridae

---

aye-aye

size	body: 14 to 17 inches (36 to 43 cm) tail: 20 inches (51 cm)
weight	4.5 pounds (2 kg)
distribution	Madagascar
habitat	rainforests; coffee, mango, and coconut plantations
diet	bamboo shoots, fruit, insect larvae, birds' eggs
reproduction	1 baby
predators	ferret
life span	23 years in captivity

class | Mammals
order | Primata
family | Dauben-toniidae

song thrush

size	8 to 9 inches (20 to 23 cm)
weight	2.3 to 3.3 ounces (65 to 94 g)
distribution	Europe, Asia
habitat	woodlands, forests, gardens, parks, meadows
diet	fruit, earthworms, insects, spiders, snails
reproduction	3 to 5 eggs; 12- to 14-day incubation period
predators	foxes, cats, mustelines, squirrels, magpies, crows, jays

class | Birds
order | Passeri-formes
family | Muscica-pidae

---

common chimpanzee

size	25 to 37 inches (64 to 94 cm)
weight	88 to 120 pounds (40 to 54 kg)
distribution	Africa
habitat	wooded savannahs, rainforests
diet	fruit, leaves, flowers, buds, bark, seeds, insects
reproduction	1 baby; 230-day gestation period
predators	leopard
life span	50 years in captivity

class | Mammals
order | Primata
family | Pongidae

---

archer fish

size	up to 10 inches (25 cm)
distribution	coasts of southern Asia and northern Australia
habitat	waters near mangrove swamps or jungles
diet	insects, small fish, crustaceans
reproduction	3,000 eggs

class | Fish
order | Perciformes
family | Toxotidae

---

panther

size	body: 3.3 to 6 feet (1 to 1.8 m) tail: 3.3 feet (1 m) height: 18 to 31 inches (46 to 79 cm)
weight	66 to 122 pounds (30 to 55 kg)
distribution	Africa, Arabia, the Far East
habitat	forests, savannahs
diet	antelopes, birds, monkeys, snakes, fish, carrion
reproduction	1 to 3 young; 90- to 112-day incubation period

class | Mammals
order | Carnivora
family | Felidae

---

koala

size	24 to 34 inches (61 to 85 cm)
weight	9 to 33 pounds (4 to 15 kg)
distribution	Australia
habitat	eucalyptus forests
diet	eucalyptus leaves and bark
reproduction	1 baby; 25- to 35-day gestation period
predators	dingoes and domestic dogs
life span	20 years in captivity

class | Mammals
order | Marsupialia
family | Phasco-larctidae

---

turkey vulture

size	25 to 30 inches (64 to 76 cm) wingspan: 6 feet (1.8 m)
weight	3 to 4.4 pounds (1.4 to 2 kg)
distribution	Canada, United States, Central America, South America
habitat	fields, meadows, roadsides, lakes, coasts
diet	carrion, young herons, ibises, fish, live insects
reproduction	1 to 3 eggs

class | Birds
order | Falconi-formes
family | Cathartidae

Glossary

baleen: Horny plates attached to the upper jaws of baleen whales

camouflage: A disguise meant to hide an animal from its enemies

carnivore: A plant or animal that feeds mainly on meat

carrion: Dead and rotting flesh

constrictor: A snake that crushes or suffocates its victim by wrapping itself around it and tightening the coils

cram: To feed, often by force, a huge amount of food

crustacean: Any of a class of animals that have shells and many pairs of legs

digest: To change food into a form that a body can absorb

groove: A channel or furrow sunk into a surface

herbivore: An animal that feeds on grass and other plants

impale: To torture or kill by piercing with something pointed

inject: To introduce a fluid into the body of an organism

insectivore: An animal or plant that feeds mainly or solely on insects (the mole is an example of an insectivore)

invertebrate: An animal without a spine

larva: An often wormlike form that is one stage of an animal's life

mammal: A member of any animal species in which the female has mammary glands for feeding her young

marsupial: An animal whose young spend several months after birth in their mother's pouch

mollusk: An animal with a soft body that has no bones but usually has a hard shell

mucus: A transparent thick liquid

nectar: A sweet liquid produced by certain plants

palate: The roof of the mouth

paralyze: To make powerless, unable to move

parasite: A living organism that lives on or in, and gets nourishment from, a host organism; often harmful to the host

pincer: A limb shaped like a vise and used for grabbing and squeezing

predator: An animal that destroys or eats another

prey: An animal that is the victim of a predator

primate: An order of mammals that includes human beings, apes, monkeys, and some related animals

rodent: A mammal such as a mouse or a hamster with front teeth made for gnawing

secretion: A liquid substance produced by the cells of a living organism

shortage: A lack or deficiency of something necessary, such as water or food

symbiotic: Living with or close to another organism, usually to help each other

tick: An arachnid that feeds on the blood of certain animals

venom: The poison that is secreted by animals such as snakes

vulture: An animal that feeds on carrion

31

ANSWERS TO THE "WHO AM I?" GAME

1.	The chimpanzee		6.	Dung beetles
2.	The oak-moth caterpillar		7.	Piranhas
3.	The crocodile		8.	The desert locust
4.	The thick-lipped bear		9.	The hairy-tailed mole
5.	The starfish		10.	The blowfly

Index

Editorial Director Caroline Fortin **Research and Editing** Martine Podesto **Documentation** Anne-Marie Brault, Anne-Marie Labrecque **Page Setup** Lucie Mc Brearty
Illustrations François Escalmel, Jocelyn Gardner **Translator** Gordon Martin **Copy Editing** Veronica Schami **Gareth Stevens editing** Joan Downing
Cover Design Joel Bucaro, Scott Krall